Cotton Culture
Bulletin No. 97

by Oklahoma Agricultural Experiment Station

with an introduction by Roger Chambers

This work contains material that was originally published in 1912.

This publication was created and published for the public benefit, utilizing public funding and is within the Public Domain.

This edition is reprinted for educational purposes and in accordance with all applicable Federal Laws.

Introduction Copyright 2018 by Roger Chambers

Self Reliance Books

Get more historic titles on animal and stock breeding, gardening and old fashioned skills by visiting us at:

http://selfreliancebooks.blogspot.com/

Introduction

I am pleased to present yet another title on Gardening.

The work is in the Public Domain and is re-printed here in accordance with Federal Laws.

As with all reprinted books of this age that are intended to perfectly reproduce the original edition, considerable pains and effort had to be undertaken to correct fading and sometimes outright damage to existing proofs of this title. At times, this task is quite monumental, requiring an almost total "rebuilding" of some pages from digital proofs of multiple copies. Despite this, imperfections still sometimes exist in the final proof and may detract from the visual appearance of the text.

I hope you enjoy reading this book as much as I enjoyed making it available to readers again.

Roger Chambers

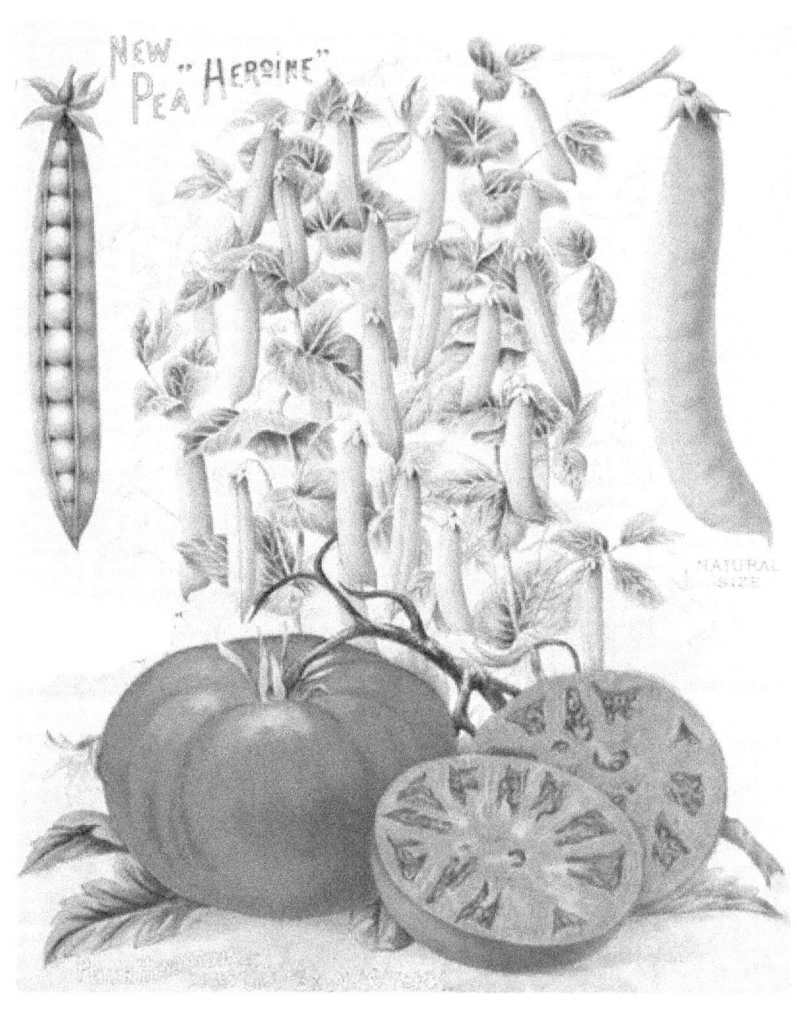

COTTON AND COTTON CULTURE

O. O. CHURCHILL AND A. H. WRIGHT

FOREWORD

We have in course of preparation an exhaustive bulletin on cotton, recording in detail all the experiments being conducted at this Station with this crop. It has been found necessary, however, to publish a bulletin prior to the time a complete treatise can be given in order to have it reach the cotton planters in time to be of use this season. This bulletin is intended to furnish the information regarding cotton and cotton culture for which the Station is most frequently called upon. A great many new citizens coming from other States and unfamiliar with the cotton crop, are constantly asking for detailed information in regard to the cultural methods, and it is hoped that this bulletin will be of special use to them.

Experimental data will be given only in the case of variety tests. The statements made, however, relative to the other phases of the subject, are based on experimental evidence. These methods are also pursued by our most successful cotton growers. It should be remembered that while certain operations should be varied according to local conditions, that agriculture is a science as well as an art, and that being a science the general principles of the different farm practices are always the same.

VARIETY TESTS WITH COTTON

Many varieties of cotton have been grown in the variety test since the organization of this Station. In order to determine the best varieties for the State, and to secure the best ones as a foundation stock for breeding, it is necessary to grow them in competition under identical conditions. The variety test is therefore carried on first, to determine the best varieties now commonly grown, and afterward to determine the relative values of the selections made in the breeding plots.

The record in 1911 in the variety test and the summary is here given:

Varieties of Cotton—1911

Variety.	Pounds per Acre.	Per Cent. of Lint.
Texas Storm Proof	526.6	36
Cook's Improved	506.7	35
Mebane's Triumph	570.0	37
Hamilton's Ounce Boll	292.0	33
Extra Big Boll	696.6	38
Bennett's Selection	483.3	33
English Ounce Boll	726.8	35
Covington's Toole	617.0	35
Virgatus	477.0	34
Ferguson's Round Nose	327.0	36
Rowden—No. 116	326.6	35
Covington's Toole	586.6	37
Mebane's Triumph	596.5	35
Texas Storm Proof	450.0	35
Cook's Improved	673.3	38
Mebane's Triumph	646.0	36
Hamilton's Ounce Boll	319.0	32
Simpkin's Prolific	762.0	34
Mebane's Triumph	465.0	37

Summary of Variety Tests of Cotton

Variety.	1903	1904	1906	1907	1909	1910	Yrs.	Average.
Parker	499.9	731.5	219	560			4	502.6
Truitt		703.5	171	546			3	473.5
Griffin	196.8	329.0					2	262.9
Allen's Improved	156.9	402.5					2	279.7
Russell		728.0	288				2	508.0
Pride of Georgia		640.5	258	640			3	512.8
Hamilton's Ounce Boll				628	800	535.5	3	654.5
Mebane's Triumph				636	590	468.2	3	564.7
English's Ounce Boll					680	355.6	2	517.8
Bennett's Triumph					680	353.4	2	516.7
Extra Big Boll					530	293.1	2	411.5
Covington's Tools					460	415.5	2	436.2
Texas Storm Proof	800.5	144.2	354	702	805	346.5	6	741.7
Cook's Improved				252	780	426.5	3	468.1

The varieties of cotton are planted on plots of uniform soil and are all given exactly the same treatment. The varieties are planted in duplicate plots. Corn has preceded the cotton in every test. Some years the corn has been cut and removed from the ground, while in other seasons a stalk cutter has been used. In every case the soil has been stirred between the middle of September and the middle of October, usually from a depth of five to eight inches, depending upon the amount of moisture present and the hardness of the soil. The soil is given no further treatment until spring.

The ground is double disked two or three weeks before seeding time. To prevent evaporation and keep the weeds in check earlier preparation may be necessary. After disking, the ground is thoroughly harrowed. If the weeds begin to grow before time to plant the cotton, the harrow is used frequently enough to keep them in check. A lister is used in preparing the seedbed, but the ridges thus formed are nearly leveled by the use of a plank drag. The cotton is then planted on top of the ridges with a two-row combined cotton and corn planter.

The cotton is chopped to the desired stand when the plants have put out from four to six leaves. In the variety work the plants are left sixteen inches apart in the row, with the rows forty inches apart. From three to five cultivations are given, depending upon the season. The first cultivation is given as deep as possible without covering the plants. The second cultivation is also deep, but the later tillage is made very shallow.

The time of planting has varied markedly, depending upon the seasonal differences. An early crop is desirable, hence the crop is planted as soon as the ground is warm enough so that the seed will germinate readily.

Much of the data secured in the variety tests have not been included in the table presented. The work has now progressed far enough so that we feel reasonably safe in drawing certain conclusions. The Big Boll Storm Proof types are the only ones which should be grown in Oklahoma. All other types have rapidly lost in favor as the result of their low production and have been dropped from the tests. Covington's Toole is the only exception to the preceding statement. The leading varieties now grown in this State, and belonging to the class just mentioned, are Mebane's Triumph, Texas Storm Proof, Cook's Improved, Ounce Boll, and Rowden.

In selecting a variety, it should be borne in mind that the variety name is not of as much importance as the selection which the seed has been given. It is better to secure a high yielding strain and one which has been given careful selection by cotton breeders than to depend entirely upon the variety name. For instance, Mebane's Triumph is one of the very best varieties of cotton for this State, yet it would be possible to find seed which has been handled carelessly for several years which would be inferior to many other varieties. The breeding and selection is of the greatest importance.

According to Duggar's classification, somewhat modified by the United States Department of Agriculture, cotton may be divided into the following types:

1. Big Boll
2. Long staple
3. Cluster
4. Semi-cluster
5. Early or short limbed
6. Long limbed
7. Peterkin or Rio Grande
8. Intermediate.

So far as Oklahoma growers are concerned, the first, or Big Boll, group is the only one of importance. Strains and varieties of the other groups, while they are all represented to some extent in this

State, are rapidly disappearing and exist now only as mixtures with the Big Boll type rather than as separate varieties. This group of cotton is described in Bulletin No. 173 of the U. S. Department of Agriculture as follows: "The essential character of the Big Boll group is the size of the boll, or to be more exact, the weight of dry seed contained in the boll. The maximum size of bolls in this group is at present about 11.5 grams, or from 38 to 40 bolls to the pound, and the minimum size has been arbitrarily fixed at 6.5 grams, or 68 bolls to the pound. The plants are stocky and usually vigorous, limbs strong and heavy, usually two in number, fruiting branches quite strong, ranging from very short or irregularly jointed or semi-cluster to very long jointed; leaves large, becoming almost glabrous; lobes broad and short, bolls large with four or five locules, seed large, fuzzy, dark green, greenish or brownish gray or white; lint short to medium, 2o to 30 milimeters in length, soft and of good strength, usually 33 1-3 per cent or more."

Figure 1.—Good Type Cotton Plant

"A subdivision of the Big Boll group has been developed on the plains west of the Mississippi, where severe rains and wind storms are frequent during the picking period. It is known as the Big Boll Storm Proof group, and includes some of the most highly developed varieties grown at the present time. The plants are vigorous, upright in growth during the first part of the season, but later drooping under the weight of the bolls. The bolls are not borne upright on the branch, but lie close to it, the peduncle, or stem of the boll, forming an angle with the branch. When the stem and fruit branches are bent down, the bolls are inclined or inverted so that when ripe the broad, thick segments of the bur and the unusually large involucres form a more or less perfect roof above the locks of seed cotton which hang down underneath and coalesce into a single mass. The locks are also securely attached to the bur, but as a rule storm proof cottons are easier to pick than varieties which have locks which more readily dislodge."

SUGGESTIONS ON METHODS OF IMPROVING COTTON

The fact that cotton can be improved to an appreciable extent and that an improved strain can be maintained has long been established by investigators and practical cotton growers. The methods suggested under this heading are not intended for the specialist in cotton breeding, but rather for the cotton grower who wishes to keep his production up to a high standard. There will always be a place for the specialist in cotton breeding, but every grower may at least keep his seed pure and of high quality even if he does not increase its productivity.

In the beginning the work of cotton improvement, it is important to secure a good strain of seed. This is best secured from some high yielding strain or variety grown in the locality in which it is intended the cotton should be grown. If the grower has on his farm a strain of cotton which has given good returns and is of the type desired, it is advisable to use it as a basis for his future work. Any standard variety of a big boll storm proof type will be good foundation stock for work in Oklahoma. Such seed should be planted and cultivated according to methods described later in this publication. The improvement will

FIGURE 2. Poor Type Cotton Plant.

then be made from individual plant selections in the field. The party making the selections should have definitely in mind what type of plant is best for his conditions and should make his selections accordingly. A study of the factors which constitute a desirable type is necessary in deciding upon the type of plant. The more important of these characteristics are: High production, early maturity, storm resistance, and high per cent of lint. In connection with this, we should also have a well formed plant with an abundance of fruiting limbs. Early maturity is not a characteristic of the large bolled group of varieties, yet by careful selection they have been made to mature sufficiently early. This characteristic is of a special importance in the northern portion of the cotton growing belts, and in weevil infested districts.

Early maturity is indicated by the position of fruit limbs on the plant. The lower limbs should be very near the ground. Storm resistance is indicated by the presence of a heavy involucre, but more particularly by the inverted or drooping position of the bolls. This

FIGURE 3. Cotton Varieties.

position will protect the boll from winds and from beating rains.

Selections should not be made from other types, such as cluster and semi-cluster types, even though desirable individual plants are available. The most convenient and satisfactory method of making selections is to go through the field before any of the cotton has been picked, and when at least 25 per cent of the bolls are open. The cotton can be picked from the selected plants at once; or, in case it is thought that enough seed cannot in this way be obtained, the selected plants may be marked in such a way that they can readily be located at the time of picking. It is desirable to gather the cotton from the selected plants before the general field picking takes place.

It is not of as great importance to select any particular picking, as many seem to believe, for the individual plant is always the unit for selection. The only exception to this would be in the case of crossing and when hybrids are produced. Care should be taken that the seed saved is kept pure, and is not mixed at the time of ginning. It should be well stored until the time for planting. If only a small amount of seed is saved, it would be well to plant this in a block by itself, and then the bulk of the seed for the coming year could be saved from the selected plot. It ought not to be hard to keep cotton seed pure and up to a high standard.

CULTURE

(Soils)

Cotton is grown on almost every type of soil found in the cotton sections of the State. The more important of these types are: Chocolate clay, heavy clay loam, fertile sandy land, upland clay, upland clay loam, and upland sandy soil. Cotton does best on sandy loam, but it may be successfully grown on any of the types of soils mentioned. Next to sandy loam, clay loam is best, and cholocate clay soil is also good. A soil containing a large amount of sand is especially advantageous in the northern part of the cotton section, for this type of soil becomes warm much earlier in the spring than do the heavier clay soils. Sandy soils are also much more easily cultivated, a better seedbed can be prepared, and as a result a better stand secured.

PREPARATION OF A SEEDBED

The time for preparing the seedbed should depend to some extent upon the previous crop grown. It should also depend greatly upon the type of soil. Clays are much more difficult to handle than are sandy soils, and much care must be used in their preparation to attain the best physical structure. When oats or some similar grain crop precedes cotton, the soil should be stirred as soon as possible after the grain is harvested. In some sections difficulty will be met in doing this. The greatest objection to this practice is that a heavy growth of weeds may be produced late in the summer, which will use a large quantity of moisture and make it necessary to retill the ground in the fall or winter. In cases of this kind it may be advisable to allow the soil to remain without plowing until a large per cent of the weed seeds have germinated.

Soil which is not broken during the summer should certainly be plowed in the late fall or early winter. When the cotton crop is preceded by corn or cotton, the stalks should be cut as soon as the crop is harvested and the ground plowed as soon as possible. This should be done before January 1. These statements regarding the time of plowing do not apply to all types of soil, as light sandy soils or soils which blow readily should not be left exposed to the winter winds. The early plowing of sandy soils is not of as much importance as early plowing of clay soils. No matter what kind of soil the cotton is planted on, it should be plowed early enough so that a good, firm seedbed may be secured.

DEPTH OF PLOWING

The depth of plowing for any crop should depend upon the type of soil and the time of plowing. On other than loose sandy soils, when plowing is done early, it should be deep. It is doubtful if it is ever advisable to plow any kind of soil for cotton deeper than eight to ten inches. In some cases deeper plowing may give slightly increased yields, providing the depth is increased gradually, but it is doubtful if the increased yield would pay for the extra expense of the deeper plowing. Where the depth of plowing has always been shallow, it is best to increase it not more than one or two inches in any one season. Spring plowing should not be as deep as fall plowing, unless sufficient work is done to firm the soil and put it in the best of tilth. Spring plowing is, however, a bad practice, except on the lighter types of soil.

It is doubtful if it will be found advisable to harrow soils immediately after plowing in the fall, especially those types which have a tendency to run together and become compact. The sticky clay upland soils, found in many parts of the State, serve as an illustration of this type. Such soils should remain unharrowed throughout the winter. This will give a better texture, and the spring preparation will be much easier.

Soils which have been prepared in the fall or winter should be thoroughly disked in the spring as soon as the weeds begin to appear above the surface of the ground. When the spring is backward it may be necessary to disk and harrow a second time before planting in order to destroy the weeds and put the soil in a proper tilth. There are two methods of seeding cotton practiced in Oklahoma—level seeding and bedding. While we have no experimental data on the subject, yet we have practiced both methods on the experimental plots. Little difference in yield has been noted between these two methods of preparing the seedbed. During the last few years we have been practicing bedding almost entirely. This method will probably be found advisable, especially in the northern part of the cotton belt where the soils are somewhat cold and are liable to be wet at the time of seeding. When bedding is practiced, the soils will warm up faster, and it is easier to secure a good stand. A little advantage is also found when it comes to cultivating the first and second times. The common lister is used in bedding the soil. The ridges are nearly leveled off by means of a plank drag which crushes the clods, compacts the soil and makes it possible to do a good, uniform job of seeding.

PLANTING

Planting should be done as soon as the soil becomes sufficiently warm so that seed will germinate readily. No definite time can be set for planting, as it should depend entirely on the climatic conditions. Many successful cotton growers prefer to seed early and run the risk of losing the first seeding rather than to wait until the season is far advanced. Perhaps personal experience will decide this question for the cotton growers. We prefer to seed a little later, rather than to run the risk of having to replant. In this State seeding is done from the first of April to the middle of May.

Cotton seed should be covered very shallow, especially if planted early. It may be planted deeper later in the season when the ground is warm, and there is not so much danger of heavy rains. When very early seeding is practiced the seed should barely be covered, but when planted late it is well to put the seed in moist soil, providing this does not require planting more than two and one-half inches deep.

Different writers recommend the planting of from one peck to one bushel per acre. In our experiments we have found one-half bushel per acre to be very nearly the proper quantity to plant.

Cotton is in nearly all cases drilled and afterward chopped to a stand. A few practice checkrowing. This method has not been given a trial in our experimental plots; hence, we cannot state positively its merits or demerits. We feel safe in saying that no advantage is gained in checkrowing cotton unless it might be on very fertile soils, rich in humus, where the cotton plants are inclined to produce an abundance of long limbs and where weeds are very difficult to eradicate by cultivation. At other Stations on such types of soil it has been recommended that the rows be planted four and one-half feet apart, leaving the spaces between the plants in the row not greater than three feet.

CULTIVATION

If a crust forms on the surface after the cotton is seeded, it is a good practice to give the field a light harrowing. The first cultivation should be given as soon as the plants are up, and in some cases it may be well to cultivate before this time. It will be found necessary to cultivate shallow the first time in order to prevent covering the little plants. The second cultivation should be as deep as it can be made without throwing too much soil upon the plants. The third cultivation, if given soon after the second is completed may be nearly as deep as the second cultivation. The succeeding cultivations should gradually decrease in depth until there results a simple pulverization of the surface soil.

The frequency of cultivation should depend a great deal on seasonal conditions. There is little danger of cultivating too often, although so far as returns obtained are concerned, such could be the case. The early cultivations should be made frequently enough to keep in check the growth of weeds until the cotton has thrown out sufficient leaves to shade the ground. The later cultivations should, as far as possible, be made after rains in order to prevent the formation of a surface crust, and in this way prevent the loss of moisture by evaporation. Four or five cultivations are usually considered sufficient. The kind of tool used in cultivating cotton is to some extent a matter of personal opinion. Our experiments at the Station have shown that there is little difference between cultivating cotton, corn and other inter-tilled crops. We use the same implement that we use in cultivating corn. The one-row, six-shovel cultivator does better work than the four-shovel cultivator.

After the cotton begins to set fruit and there is danger of injuring the plants by knocking off the squares and bolls if the two-horse cultivator is used, we find it advisable to use the one-horse, five-tooth cultivator. In very dry seasons, when the main purpose in cultivating is to conserve the moisture, we have found it very valuable to drag a mower wheel between the rows. The one-horse "A" harrow, made by driving spikes through the planks, is also a very valuable implement. The important thing, of course, is to keep the land in the right condition, and it matters little what kind of an implement is used so long as the right results are secured.

COTTON ROTATIONS

In order to be of the greatest value, a rotation must be more than a mere change of crops. It should be so planned that it will assist in maintaining the fertility of the soil, in maintaining a good physical condition of the soil, in checking injurious insects and weeds, and in providing a more economical arrangement of farm work. A rotation which will assist in maintaining the fertility will, in a large measure, possess the features necessary to provide the other requirements of a rotation. To devise such a rotation is of chief importance.

It will be found necessary to include in the rotation some crop that will return to the soil nitrogen and organic matter. Any legume will serve this purpose, but cowpeas will be found much more valuable than any now commonly grown. This crop can be grown as a spring crop or as a catch crop after wheat, oats or corn. If grown as a catch crop, cowpeas should be sown as quickly as possible after the grain crop is harvested. The vines should be plowed under in the fall. When cowpeas are grown as a regular crop, the greatest returns will probably be secured when most of the pods are gathered before the crop is plowed under. If cowpeas are harvested and all parts removed from the soil except the roots and stubble, no increase in fertility will occur, but the drain on the soil nitrogen will be much less than when a grain crop is grown. However, the physical condition of the soil will be greatly improved.

Cotton has been grown in several different rotations on the experimental plots, and in a field way on the College Farm. From these experiments the following rotations are suggested for different portions of the State, but it should be remembered that the ones given may not always be the best under all conditions:

Three-Year Rotation

First year, corn.
Second year, oats followed with cowpeas.
Third year, cotton.

First year, kafir corn.
Second year, wheat followed by cowpeas.
Third year, cotton.

Four-Year Rotation

First year, cotton.
Second year, oats followed with cowpeas.
Third year, cotton.
Fourth year, corn.

First year, cotton.
Second year, oats.
Third year, cowpeas.
Fourth year, corn.

These rotations are suggested as among the most successful on certain kinds of farms. The rotation, of course, should be modified to suit the kind of farming being practiced. It should depend upon the number and kind of live stock kept, and upon which crops are made the leading market ones. In some seasons the cowpeas are not a success as a catch crop after oats or wheat, due to the very dry summers which we sometimes experience. The price of cowpea seed frequently keeps farmers from growing them merely for green manuring purposes. This objection can be largely overcome by growing a sufficient amount of seed on the farm where it is to be planted.

FERTILIZING COTTON

The experiments with commercial or mineral fertilizers have not progressed far enough at the present time to draw positive conclusions. Barnyard manure has been found very valuable in the cotton rotation. The yield of seed cotton has been increased from forty to one hundred pounds by the use of manure, above that secured from plots not manured. The effects of the manure can be noted in the yield for several years after the application has been made. The following statements concerning this subject are quoted from Oklahoma Station Bulletin No. 77: "With the exception of cases where it is expedient to burn the stalks, they should be cut in short pieces with a stalk cutter and incorporated with the soil. In sections where the cotton boll weevil has made its appearance, burning all refuse upon the field has been recommended for the purpose of destroying as many of these injurious insects as possible. It should be remembered that the process of burning opens an avenue for the escape of that important element, nitrogen, and organic matter is destroyed. Hence, as a matter of fact, the store of substances which add materially in securing perfect physical structure in the soil is diminished. If we compare the composition of the seed with the chemical elements found in the lint, it will be seen that the latter contains an insignificant amount of the fertilizing constituents. Thus, our inference that cotton is not an exhaustive crop is correct, providing the fiber alone becomes an article of commerce. If this statement holds true in farm practice, the seed should be fed to the live stock on the farm, and the well preserved manure must find its way back to the field from which the original crop was taken. There is another fact which is pertinent to this subject, but is not brought to our attention in the analytical work referred to. A fourth division of the cotton seed may be made; thus, a second product, cotton oil, can be placed upon the market, and the minimum quantity of the essential plant food element will be extracted. Cotton oil, like the lint, contains a very small amount of nitrogen, phosphorus and potassium. If the seed is sold from the farm and is subjected to the manufacturing processes of the cotton oil mill, the cotton seed meal can still be utilized for feeding purposes, thereby giving the grower a chance to maintain the soil fertility."

INSECTS

The following notes on insects and methods of controlling them were prepared by Professor C. E. Sanborn, of the Department of Entomology:

There are many different kinds of insects which are particularly injurious to the cotton plant. These do not attack the plant at any certain time in the growth of the plant, but one or another of the species may cause a continual infestation from the time the seeds become well germinated until the plant has matured.

Sometimes the infestation is so meager that it is scarcely noticeable, then again it may be so severe as to destroy the entire crop. The tax on the vitality of a plant is therefore great, and the final tax on the entire State crop figured in dollars amounts to above a million. Thorough cultivation of proper varieties of cotton for Oklahoma aids in making a profit, but methods of culture and seed selection will never prove entirely satisfactory for controlling these ever-present pests.

The cut worms are the first to attack the crop. These pass the winter as nearly full grown caterpillars, especially in fields that have been in grass or were exceedingly foul the previous year. The most conspicuous cut worm work is that of cutting off the young plants near the surface of the ground. A less conspicuous type of infestation is the defoliation of the plants. In either case the work is generally done at night, so that only signs of the insect are apparent during the day.

Remedies: In cases where the foliage is being devoured, dust the plants with paris green at the rate of one pound to the acre. In cases where the plants are cut off near the ground, use poison bran mash, which is made as follows: Mix 1 pound of paris green in 25 pounds of wheat bran. Make of this a stiff mash by adding slightly salty water, a little at a time, while stirring the bran and poison. Broadcast this on the infested places of the field in the evening. The worms on coming out of their hiding places in the soil will eat of this if it is conveniently placed over them, and they will cease to eat of the plants.

The Cotton Louse.—About as soon as the seeds become well germinated the lice begin to infest the underside of the leaves, causing them to become unhealthy, and a consequent stunting of the plant soon follows at the very season of the year when a rapid growth is most desirable. If at this time the weather remains cloudy and damp the lice will become exceedingly abundant, because they produce young alive, and their natural enemies, such as parasites which develop from eggs, cannot hatch unless the weather is dry and warm.

The lice can be controlled by the application of a spray from beneath the leaves upward, of Black Leaf diluted with water at the rate of 1 gallon of the former to 60 gallons of water, or by the application of a concentrated form of Black Leaf, known as Black Leaf 40. One gallon of the latter will make a thousand gallons of spray fluid when properly diluted.

It is not generally advisable to spray cotton on account of the cotton louse, except in cases of intensive cultivation where the greatest yield possible is desired.

The Careless Weed Worm, also known as Alfalfa Web, is about the next to appear. It usually infests the field in spots or patches, but sometimes works rather uniformly over a field. The worm works much the same on cotton as alfalfa, defoliates the plant to a great extent and binds some of the leaves together for its protection by means of its web. There are three or four broods of this insect per year, but one, however, usually the first, attacks cotton when the latter is about six or eight inches high.

Remedy: Dust the infested localities in the field as soon as the worms appear with paris green or lead arsenate at the rate of 1 pound per acre. If considerable damage is done by the time the worms are noticed, dust with lead arsenate at the rate of 2 pounds per acre, or in case of a liquid spray use 4 pounds of lead arsenate to 50 gallons of water.

Grasshoppers are often very injurious to cotton. Caution should be used and as soon as an infestation is observed moving toward or in the cotton field, apply a dust spray of paris green or lead arsenate at the rate of 1 pound per acre. If the grasshoppers must cross a grass or weed patch, make a poison guard by dusting said vegetation with paris green at the rate of about 2 pounds per acre.

The Boll Worm.—Although easily controlled when infesting nearly any plant except corn, this insect is responsible to the Oklahoma cotton farmers for a loss of thousands of dollars. There are three or four broods of this pest each year. The first brood appears when corn is a little less than knee high, the second when sweet corn is in roasting ear, and the third about the first of August, or just after the time that field corn is ordinarily maturing. This third brood is the most injurious of any to cotton, and unless proper precaution is taken the eggs laid on the leaves and squares of cotton will develop and do considerable damage.

Remedy: Just as soon as the moths are known to be laying eggs on the cotton, dust it with 1 pound of paris green or lead arsenate to the acre, if less than knee high; if higher, use from 1½ to 2 pounds of the poison. See to it that the first mouthful of the young boll worm is its last. This can be done, and should be done, by all cotton raisers.

Remedy for general control of the Boll Worm: The best general remedy for preventing the depredations of the boll worm on corn, cowpeas and cotton is given in the following brief sentence: **Plow infested fields deeply in the fall.** The boll worm, like the white grub, cut worm, corn hill bug and many other injurious pests, passes the winter in the ground. Such a disturbance as is brought about by fall plowing breaks up its house and only a small per cent, such as escape displacement, are enabled to live through the winter.

The Cotton Leaf Caterpillar.—This insect is often mistaken for the boll worm. It does not ordinarily injure the bolls, but devours the leaves and squares, if it is not properly controlled. The plants should be treated for this insect the same as for the boll worm.

The Boll Weevil.—This bug is present only in the southeastern part of this State. If all the planters in the infested localitiy could act in unison by planting no cotton for one year, or planting only an early maturing variety and destroying the cotton stalks at least one month before the appearance of the first frost in the fall, the insect would practically be eliminated because cotton is the only food plant. In order for it to be able to withstand the winter conditions of Oklahoma, it must have green cotton and eat right up to the appearance of frost.

The destruction of cotton stalks by deep plowing is advised. If for any reason this is not practicable in any part of the boll weevil infested territory, another method, that of the uprooting and burning of the plants, is advised.

CLIMATIC CONDITIONS IN OKLAHOMA

Oklahoma has a great variety of climatic and soil conditions. In treating any crop for this State it is necessary to keep this point in mind. These conditions should to a large extent determine what crops to grow in the different localities. Perhaps no other State has so many different conditions entering into the production of crops.

Great differences in rainfall exist. The average precipitation varies from less than fifteen inches in the extreme northwestern corner to over fifty-five inches in the extreme southeastern corner. Dividing the State into belts of ten-inch variations we have an eastern belt with an average annual rainfall of over forty inches; a central belt with an average between thirty and forty inches; a western belt with an average of twenty to thirty inches, and a Panhandle belt with less than twenty inches. These belts often overlap and vary considerably from year to year. Taken as an average for several years, however, they can be considered fairly well established.

Throughout the year the monthly precipitation is highest in May and lowest in December and January. The monthly precipitation averages much higher in May than in any other month at most of the Stations where weather records are kept. The June rainfall is also much above the average for the entire year. Over 75 per cent of the precipitation occurs during the growing season; that is, between March 1 and October 31.

A proper distribution of rainfall is a very desirable feature in the production of crops. The records are presented in charts at the end of the Bulletin. Statistics for these graphics were secured from the Station records for Stillwater and from the reports of the U. S. Weather Bureau for all other points. The averages given are those computed by the Weather Bureau. With the exception of Kenton, Mangum, Stillwater and McAlester, the records include the precipitation to 1909, but for these Stations the records are made complete, including the records of 1911.

The following table shows the average rainfall from 1901 to 1911, inclusive, at some of the representative points in the different belts mentioned above:

MEAN AVERAGE RAINFALL FOR ELEVEN YEARS BY MONTHS

Place	January	February	March	April	May	June	July	August	September	October	November	December	Annual
Kenton, Oklahoma	0.25	0.61	0.57	1.65	2.61	1.75	2.24	2.05	1.73	0.95	0.63	0.35	15.16
Mangum, Oklahoma	0.86	1.25	0.87	2.33	4.76	4.24	2.65	2.40	2.71	2.12	1.65	0.26	26.86
Stillwater, Oklahoma	1.15	1.41	2.25	3.40	5.85	3.72	3.37	3.65	3.32	2.75	2.06	1.18	34.00
McAlester, Oklahoma	2.03	1.86	3.44	6.06	6.62	4.90	3.62	3.19	3.29	3.79	2.68	2.45	42.42

DROUTH

Oklahoma lies between longitude 94 and 103 west. No other section of the United States situated in the same longitude has as great an average rainfall as has Oklahoma with the exception of eastern Texas. The western half of the State, with the exception of Cimarron and Beaver Counties, has as great an average rainfall as Minnesota and greater than that of either of the Dakotas, Nebraska or Kansas. The eastern half of the State has an average rainfall equal to that of Wisconsin, Michigan, northern Illinois, Iowa, Missouri and eastern Kansas. The extreme counties of eastern and southeastern Oklahoma can be grouped with the New England States, the Virginias, Kentucky and Missouri.

It will be seen from these records that Oklahoma compares very favorably, as far as precipitation is concerned, with the most productive States of this country. Our inquiry is at once directed to the fact that with the exception of Texas and southern Kansas, Oklahoma suffers perhaps more from drouth than do other States having the same amount of rainfall. This being the case, it would be well for us to make a study of the socalled drouths in Oklahoma and briefly discuss methods whereby their severity can be reduced to a minimum.

The climate has a vast influence upon a region when it comes to its value for agricultural purposes. Perhaps there is no place in the United States where more profitable crops could not be produced if the precipitation came at more opportune times. All sections suffer from drouths at times. This is true in sections having an ample rainfall just as well as in those sections not so favorably located in this respect.

The Weather Bureau records show that the average annual rainfall for Oklahoma in all sections, except the Panhandle counties, is ample to produce maximum crops every season. Insufficient precipitation is not the cause of drouth in this State. Neither is drouth caused by an unfortunate distribution of the rainfall, for the records show that in all parts of the State 75 per cent of the rainfall occurs during the growing season—March 1 to October 31.

The greatest monthly rainfall occurs during April, May and June. The following reasons will account to some extent for the conditions under consideration:

1. That of evaporation of moisture from the soil caused by hot winds, prolonged high winds, and high temperature.

2. Rapid evaporation of moisture from plants due to the effect of hot winds.

3. Rapid surface drainage caused by the physiography of the land, the type of soil and an insufficient supply of humus.

Hot winds are closely associated with what is known as drouths. In this State we have suffered from drouths without experiencing hot winds, which proves that hot winds alone are not responsible for the damage. This was true in the season of 1910--11, when the wheat crop was almost a total failure in many parts of the State, and yet this crop ripens before hot winds occur. We must attribute much of the damage done to our crops to causes other than hot winds.

Drouths can be overcome to a large extent by proper preparation of the soil if the crop is given good tillage. Crops adapted to the parts of the State in which they are grown should be planted. Cotton is often checked by dry weather, but is seldom destroyed. It is never a complete failure as a result of drouths.

The severity of drouths can be reduced by so handling the soil that it will absorb a larger per cent of the rainfall, and that it will retain the moisture absorbed.

To assist the soil in absorbing the rainwater two things are important:

1. Early and deep preparation of the seedbed.
2. Adding vegetable matter to the soil.

Early and deep plowing may not be advisable where the soil is very sandy, for such soils very readily absorb moisture, and when plowed early often drift badly. The depth of plowing and the time of plowing should depend very largely upon the type of soil. The addition of humus or vegetable matter to the soil is of the greatest importance. The average Oklahoma soil is very deficient in this material. A soil well stocked with humus will hold more moisture and will not become eroded as easily as one that is deficient in this respect.

Perhaps the easiest way in this State to add vegetable matter to the soil is to plow under a crop of cowpeas. All vegetable matter should be saved and applied in some form. It is poor policy to burn corn or cotton stalks unless by so doing large quantities of insects are destroyed.

To overcome the effects of drouth the first thing is deep, early fall plowing on soils which do not drift, in order to make a reservoir for the water and then follow with proper methods of cultivation to keep the water from evaporating. The land should be well filled with vegetable matter at all times, and no weeds or foreign crop should be allowed to grow on the field. A good system of crop rotation should be introduced, as it is much easier to keep the soil in good condition and free of insects and plant diseases than when a single crop is grown. The bad effects of drouth will be largely overcome providing precautions are taken along the lines suggested.

Erratum: The legend in the central part of Figure 6, "Last spring killing frost", should read, "April 1-10", instead of "April 1-20".

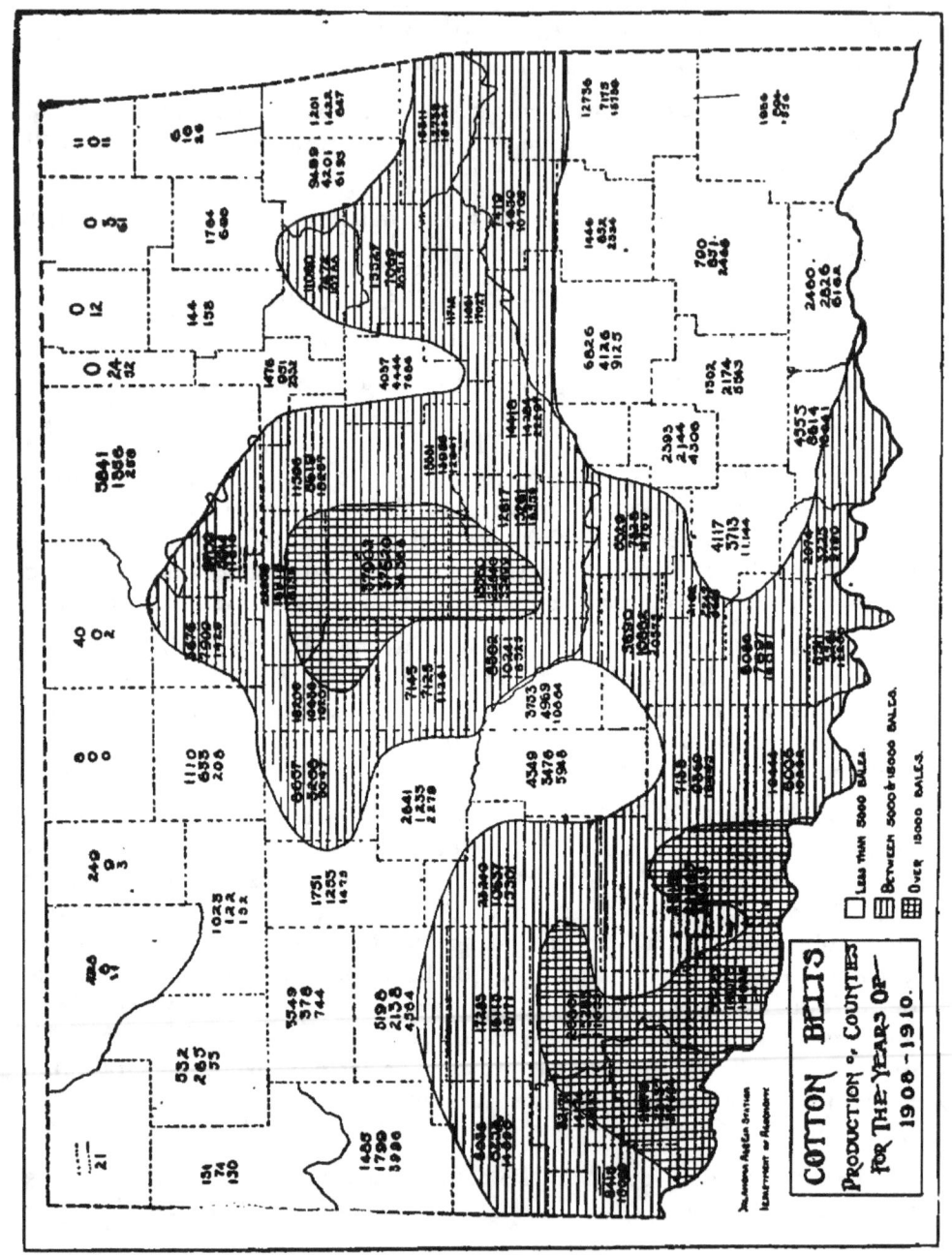

FIGURE 4

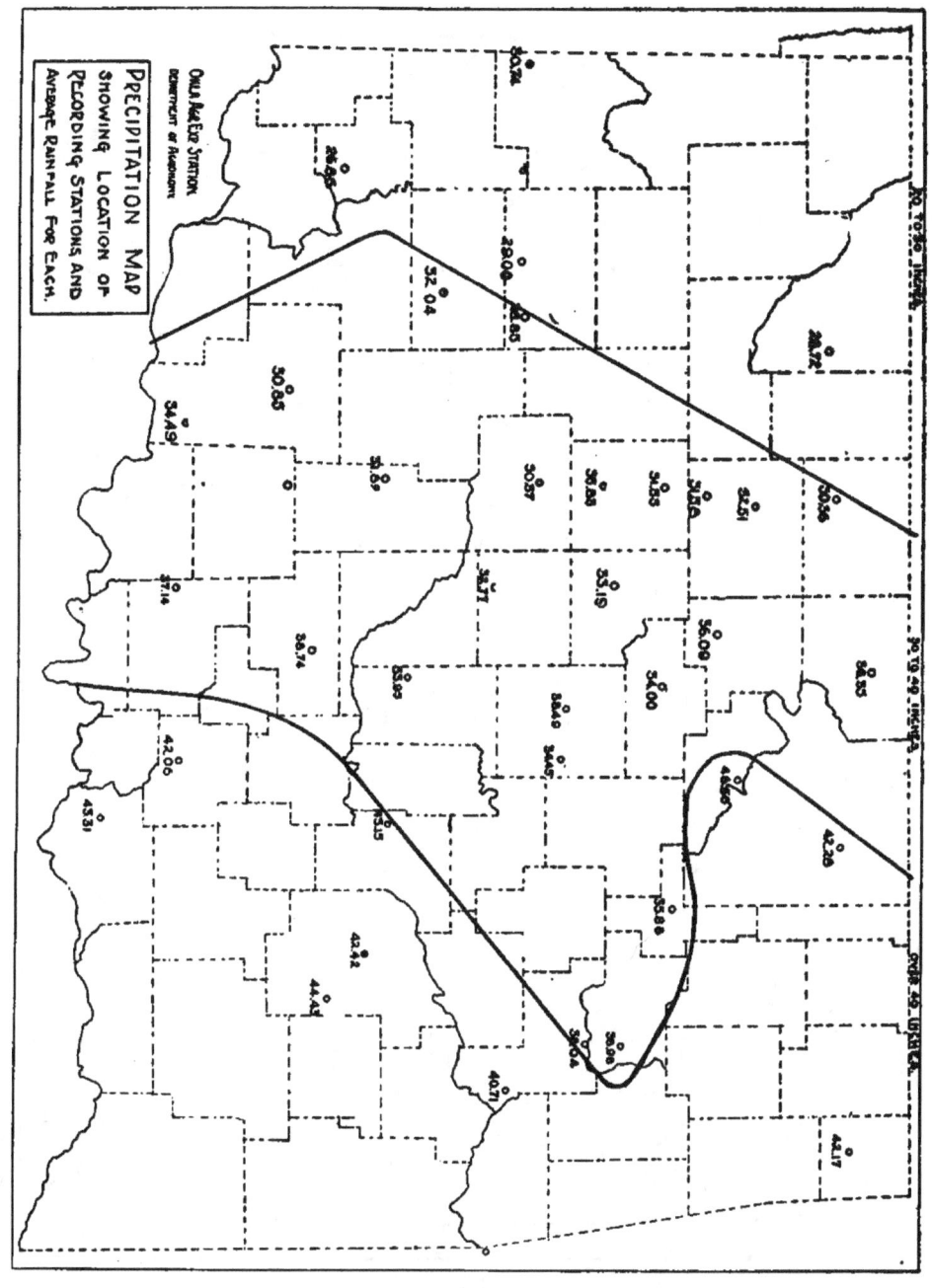

FIGURE 5

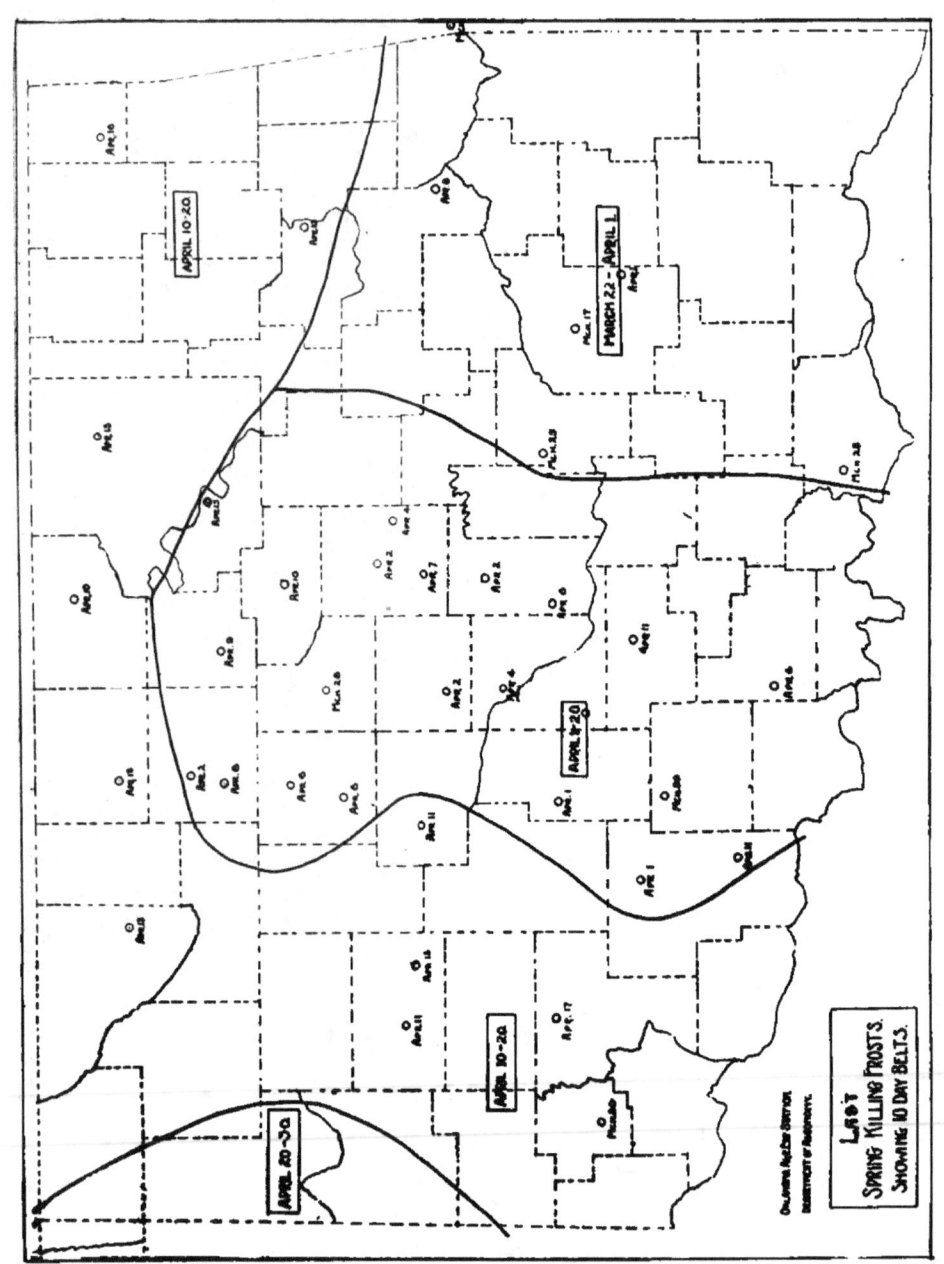

FIGURE 6

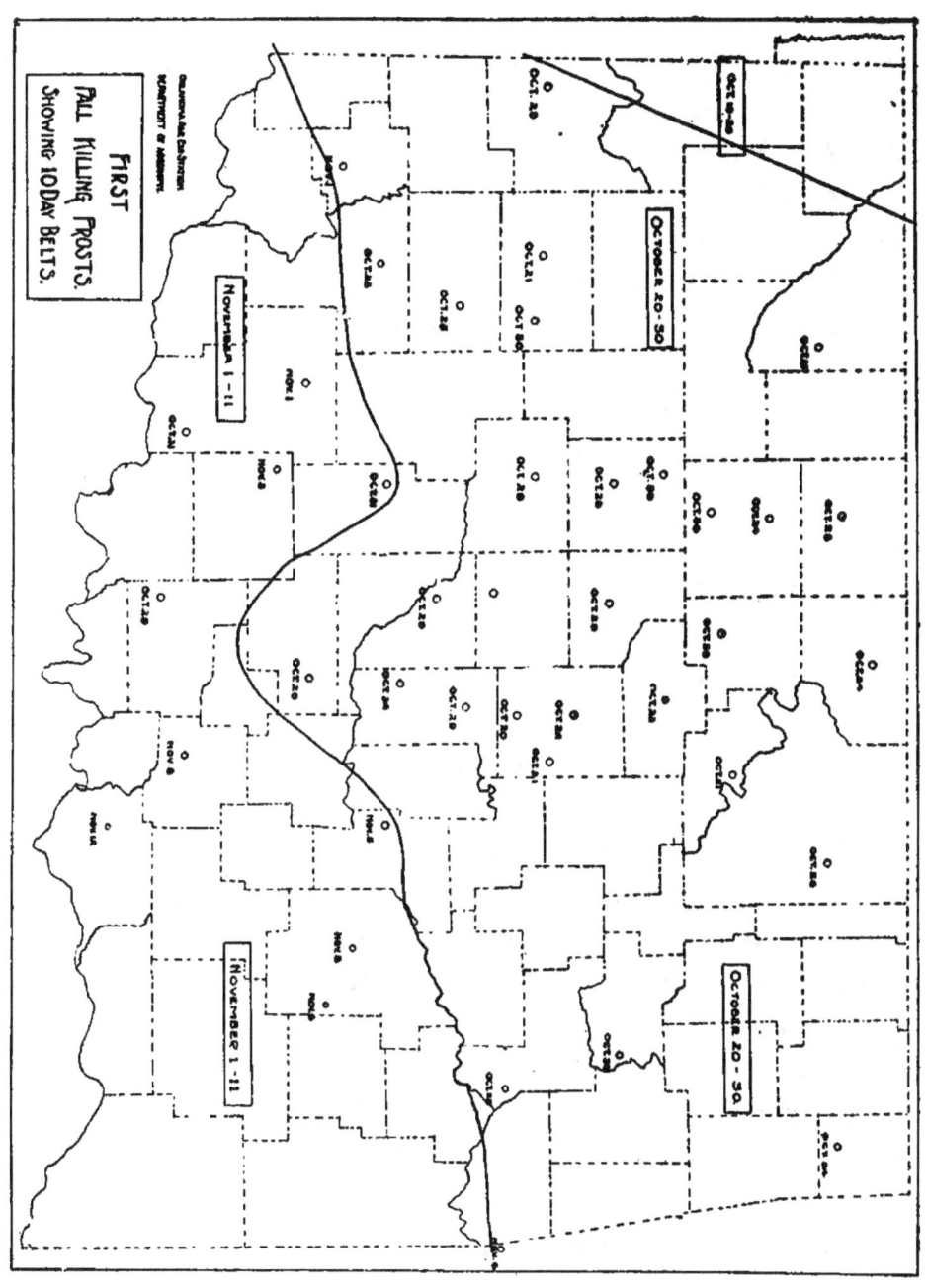

FIGURE 7

OKLAHOMA AGRICULTURAL EXPERIMENT STATION

Stillwater

The following are available publications of the Oklahoma Agricultural Experiment Station:

No. 66—The Water Supply.

No. 67—Miscellaneous Water Analyses.

No. 69—Small Fruits.

No. 72—Tests of Dips as Lice and Tick Killers.

No. 75—A Study of the Bacterial Content of Cream.

No. 87—Corn Culture.

No. 88—Southern Plum Aphis.

No. 89—Chemistry of the Kafir Corn Kernel.

No. 90—A Study of Bermuda Grass.

No. 91—The Twig Girdler.

No. 92—Spray Calendar.

No. 93—Artificial Insemination.

No. 94—Hog Feeding.

No. 95—Varieties of Fruits Raised in Oklahoma.

No. 96—Vitality of Reproductive Cells.

Circular No. 6—The Bactericidal Properties of Various Disinfectants.

Circular No. 7—The Value of Cotton Improvement.

Circular No. 12—Summary of Experiment Station Work.

Circular No. 13—Selecting an Orchard Site.

Circular No. 14—Protecting Trees From Rabbits.

Circular No. 15—Some Types of Silos.

Fourteenth Annual Report.

Fifteenth Annual Report.

Sixteenth Annual Report.

Eighteenth Annual Report.

Nineteenth and Twentieth Annual Report.